KEY PRACTICES FOR IMPLEMENTING GEOSPATIAL TECHNOLOGIES FOR A PLANNING AND ENVIRONMENT LINKAGES (PEL) APPROACH

July 2008

Prepared for:
Office of Interstate and Border Planning
Federal Highway Administration
U.S. Department of Transportation

Prepared by:
Planning and Policy Analysis Division
John A. Volpe National Transportation Systems Center
Research and Innovative Technology Administration
U.S. Department of Transportation

Contents

I. Background 1
Purpose 1
Linking Environmental and Transportation Planning Through GIS 1
Cases Selected for Study 2

II. Case Studies 3
Colorado – GeoMap2 3
North Carolina – NC OneMap 7
Oregon – Portland Metro Regional Land Information System 11

III. Case Study Findings and Conclusions 15
Findings 15
Challenges 17
Lessons Learned 17

IV. Conclusions 19

V. Appendix A: Interview Questions 20

REPORT NOTES AND ACKNOWLEDGEMENTS

This report was prepared by the U.S. Department of Transportation John A. Volpe National Transportation Systems Center in Cambridge, Massachusetts. Authors of the report were Alisa Zlotoff and Carson Poe, both of the Planning and Policy Analysis Division, and Amy Sheridan of Chenega Advanced Solutions and Engineering (CASE), LLC.

Numerous staff from several transportation agencies helped to develop this report. The authors wish to thank these individuals for their time, knowledge, and guidance.

I. BACKGROUND

PURPOSE

This report presents three case studies that illustrate how geographic information systems (GIS) have been used to implement the Federal Highway Administration's (FHWA) Planning and Environment Linkages (PEL) approach. The PEL approach provides information and tools to help agencies integrate consideration of environmental factors into transportation planning. PEL represents an approach to transportation decision-making that considers environmental, community, and economic goals early in the planning stage and carries them through project development, design, and construction. This can lead to a seamless decision-making process that minimizes duplication of effort, promotes environmental stewardship, and reduces delays in project implementation.

This report identifies some effective uses of GIS to support the goals of PEL and will be helpful to transportation and resource agencies considering applying GIS to implement PEL.

LINKING ENVIRONMENTAL AND TRANSPORTATION PLANNING THROUGH GIS

In October 2007, FHWA's Office of Environment, Planning and Realty convened a one-and-a-half day peer exchange for state Department of Transportation (DOT) GIS, planning, and environmental staff to explore how GIS might help their agencies accomplish the goals of FHWA's PEL initiative.[1] This report, highlighting how specific agencies have used GIS to meet PEL goals, is a follow-up to that peer exchange.

Along with promoting early involvement of environmental considerations in order to streamline transportation decision-making, the PEL approach encourages transportation and resource agencies to:

- consider a variety of environmental, community, and economic factors early in the planning stage,
- consider these factors throughout project development, permitting, design, and construction, and
- promote greater communication between transportation and resource agencies throughout the planning process and after project completion.

Using the PEL approach, transportation agencies can work closely with resource agencies to develop more environmentally sensitive transportation projects, avoid costly delays due to unexpected disagreements late in a project's development, support resource agency conservation efforts, and improve relationships among project stakeholders.

Recently, FHWA has looked at GIS as a tool to visually and powerfully integrate transportation and environmental data, directly serving the goals of PEL. GIS, a collection of software, hardware, and data used to store, manipulate, analyze, and present geographically referenced information, can facilitate both data-collection and analysis. Using GIS, a transportation planner can overlay location-specific data to identify areas where transportation projects might conflict with environmental or demographic factors. GIS applications also have the potential for inter-agency use, facilitating communication and information exchange. Finally, the data embedded in GIS can be made available for both transportation and land use modeling.

---

[1] A summary report of the peer exchange held in Portland, OR in October 2007 can be found at, www.gis.fhwa.dot.gov/documents/PeerEx_Report_112607.pdf

As a key step towards identifying and disseminating effective uses of GIS applications for PEL, the U.S. DOT Volpe Center (Volpe Center), in coordination with FHWA's Office of Interstate and Border Planning, developed the three case studies presented here.

CASES SELECTED FOR STUDY

The case studies selected represent three different agency types—state DOT, statewide data clearinghouse, and Metropolitan Planning Organization (MPO)— and three distinct approaches to integrating environmental and land use considerations into transportation planning.

These cases were selected through a multi-step process. First, the Volpe Center project team conducted an online literature review of GIS applications used for transportation planning purposes. Approximately 12 candidate GIS applications were evaluated on three factors: the number of data layers made available, the degree of data-sharing among the transportation agency and other agencies, and whether transportation agencies were able to access the data for land use modeling.

After selecting three cases, a series of 30-60 minute telephone interviews were conducted with several individuals from each agency, including staff from state DOTs, MPOs, and geospatial clearinghouses or organizing councils. Detailed information on each application's purpose and background was obtained through these telephone interviews. The case studies, which are included in Chapter II of this report, are based on these interviews.

The three cases are:

- **GEOMAP2 (COLORADO)** – GeoMap2, which the Colorado DOT (CDOT) hosts, is being designed to integrate environmental data from an earlier, internal CDOT application as well as a database that Colorado's North Front Range Metropolitan Planning Organization (NFRMPO), CDOT, and other partners designed. The database was originally part of NFRMPO's Strategic Transportation, Environmental and Planning Process for Urban Places (STEP UP), a pilot GIS program. GeoMap2 will be used internally at CDOT for transportation planning and project development.

- **NC ONEMAP (NORTH CAROLINA)** – NC OneMap is a state geospatial data clearinghouse hosted by the North Carolina Geographic Information Coordinating Council (GICC). Multiple state agencies, including North Carolina DOT, use NC OneMap to share environmental and infrastructure data.

- **REGIONAL LAND INFORMATION SYSTEM (RLIS) (OREGON)** – The Portland Metro MPO (Portland Metro), an organization serving 25 cities in the Portland region, hosts RLIS. RLIS is shared with Metro's member jurisdictions for land use and transportation planning. Detailed data are made available on a fee-for-service basis to consultants and developers, and a mapping function is available to the public for free via the internet.

II. CASE STUDIES

COLORADO – GEOMAP2
Website being developed

CONTACTS
Beth Baily, CDOT
Kim Hubble, CDOT
Aaron Willis, CDOT
Tracy MacDonald, CDOT
Lou Henefeld, CDOT

APPLICATION HOSTED BY:
Colorado Department of Transportation (CDOT)

KEY POINTS RELATED TO PEL

- Colorado's GeoMap2 is a web-based geospatial application providing transportation and environmental data layers.
- GeoMap2 will replace a current application, GeoMap1, which was developed for basic mapping functionalities as well as for viewing and querying geospatial information.
- GeoMap2 will expand on the capabilities of GeoMap1 by providing an interface with enterprise resource planning software to view geospatial information as users enter and update CDOT projects.
- GeoMap2 is anticipated to be complete and available to CDOT employees in the fall of 2008.

DEVELOPMENT OF GEOMAP2

In November 2006, CDOT developed an application called "GeoMap1" that allowed internal DOT users access to limited mapping functionalities through a GIS mapping interface. The interface allowed CDOT users to enter and update construction projects with geographic locations and data such as county, commission districts, and other structures associated with the selected location.

GeoMap2 is an upgrade of the original system that is currently under development and is expected to be running in the fall of 2008. It will display a range of environmental data for transportation planning and provide enhanced mapping capabilities. Internal DOT users will have access to GeoMap2 via a VPN account.

To complete the interface redesign, CDOT is concurrently integrating GeoMap2 with the Strategic Transportation, Environmental and Planning Process for Urbanizing Places (STEP UP) system.[2] STEP UP, a GIS-based environmental streamlining pilot project that CDOT, the Federal Highway Administration (FHWA), the Environmental Protection Agency (EPA), and the North Front Range Metropolitan Planning Organization (NFRMPO) in Colorado partnered to implement, was designed to improve transportation, land use, and environmental planning. A critical component of the pilot STEP UP was a GIS database that contained a wide range of environmental data,

---
[2]More information about STEP UP is available at
www.itre.ncsu.edu/ADC10/PDFs/SummerWorkshop06/Environmental_Streamlining_STEP_UP_Program.pdf and in FHWA's Environmental Toolkit State Environmental Streamlining and Stewardship Practices Database at:
www.environment.fhwa.dot.gov/strmlng/searchresults.asp?keyword=&StateSelect=Colorado&CategorySelect=all&startrow=1&ResultsSelect=10

such as parks, historic areas, wetlands, threatened and endangered species, habitat, waters of the US, and floodplains, which users could draw on to expedite transportation planning processes. The pilot was completed as a two-phase project with the first phase finishing in 2004 and the second phase in 2007.

After completion of the STEP UP pilot, CDOT and STEP UP partners collaborated to facilitate statewide expansion of STEP UP. CDOT sought to integrate the STEP UP's GIS interface with its own developing GeoMap2 program. In addition to combining STEP UP functionalities and features with GeoMap2, CDOT is now using data originally developed by CDOT's Information Management Branch to display environmental data layers for transportation corridors. GeoMap2 will maintain the user commenting functionality provided in the original STEP UP pilot.

GeoMap2 is being developed as a collaborative effort between CDOT's Planning, Environmental, GIS, and IT Sections. Ultimately, GeoMap2 will replace CDOT's GeoMap 1 and CDOT's current desktop application, Maps2, to provide broader functionality for planners and others who require environmental data for transportation planning.

ASSESSING GEOMAP2 FOR PEL

GeoMap2's integration with STEP UP is relevant to PEL for several reasons. First, STEP UP provided one model for implementing PEL. STEP UP allowed users to access critical environmental information, facilitating early collaboration with planning and environmental resource and regulatory agencies. In addition, according to the STEP UP Phase I report completed in May 2005, STEP UP was designed to:

> further strengthen the process by which projects are screened and prioritized for inclusion in regionally significant corridors by allowing both project planners and [NFR]MPO staff to review the potential environmental conflicts for corridors and individual projects. The [NFR] MPO will be able to use this information in its prioritization and screening process so that the new RTP [Regional Transportation Plan] will prioritize those projects that will avoid constrained resources and not require mitigation.[3]

STEP UP developed a GIS interface specifically to emphasize the consideration of environmental factors throughout the transportation planning process.[4] For example, users could choose specific environmental GIS layers to evaluate environmental impacts of a transportation corridor, identify cumulative and environmental issues that might affect a project's feature, and allowed users to more easily assess environmental effects for National Environmental Policy Act (NEPA) documentation.

In integrating GeoMap2 with STEP UP frameworks and functionalities, CDOT is developing a new model for how GIS can be used to support the PEL initiative. That model is based on STEP UP's presentation of environmental data using an intuitive GIS interface, but also involves integrating environmental data from across the state rather than from just one MPO. The GeoMap2 model will ultimately allow access to environmental data for internal CDOT users, while the pilot version of STEP UP allowed access only to pilot participants.

Success Factors

STEP UP project developers agreed that STEP UP met several measures of success, including increasing collaboration between decision-makers and allowing stakeholders to learn from each other. These success factors demonstrate that STEP UP has broad applications to PEL in emphasizing early collaboration among transportation and resource agencies to streamline

---

[3] The STEP UP Phase I Report is available at: www.nfrmpo.org/pdfs/PhaseIReport_v4.pdf
[4] Static screen shots of STEP UP's GIS interface can be viewed at:
www.itre.ncsu.edu/ADC10/PDFs/SummerWorkshop06/Environmental_Streamlining_STEP_UP_Program.pdf

decision-making. While GeoMap2 is still in development, initial feedback solicited at a recent brown bag seminar to demonstrate GeoMap2 has been positive, especially regarding the system's intuitive mapping functionality.

GeoMap2 staff noted an initial lesson learned:

Early planning – It was important for GeoMap2 staff to plan ahead for advancing technology. Because GeoMap2 is the first CDOT geospatial application that will be available as a ".net" application, staff had to negotiate several issues that had not previously been addressed. CDOT's Information Technology Office (ITO) has never before supported either ArcGIS Server or ".net" applications. Staff had to find contractors with the appropriate technological knowledge and ensure that CDOT's ITO was on board with the new technology.

Challenges

Staff for STEP UP and GeoMap2 have encountered several challenges when developing these systems. For example, when developing the STEP UP pilot version, participants found it difficult to identify the priority components of the system that all stakeholders could utilize. In addition, the STEP UP Phase II report identified several challenges regarding statewide implementation of the pilot.[5] Determining who should post and maintain STEP UP data was a difficult issue as well as identifying the appropriate data sources. STEP UP participants could readily contribute to a pilot version, but the process for involving participants at a statewide level was likely to be more complex. A final challenge to statewide implementation related to PEL goals was determining how to integrate all environmental review steps into STEP UP for different users, not all of whom complete environmental reviews for transportation planning in the same way.

GeoMap2 seeks to address these challenges by incorporating both environmental and demographic information into a GIS interface; when GeoMap2 is complete, users across the state will be able to easily access this information to respond to the full range of NEPA as well as SAFETEA-LU requirements for transportation planning. Furthermore, CDOT is collaborating with MPOs to share the data specific to each MPO's corridors and to coordinate policies for data-collection and maintenance.

Ongoing challenges regarding GeoMap2 development are related to questions about level of data detail. For example, both NEPA-required data and state MPO data are at the parcel level, but data from Transportation Management Areas (TMAs), federally-designated regions with a population over 200,000, are at a broader scale. Determining the appropriate level of data detail for GeoMap2 is a necessary but difficult question.

FEATURES OF GEOMAP2

Data for GeoMap2 were derived from a variety of sources, including CDOT's highway data. Environmental layers on GeoMap2 were drawn from Maps2. Data layers that will be available in the system include public lands, historic areas, wetlands, threatened & endangered species, habitat, floodplains, and hazardous waste.

Currently, GeoMap2 will be available only to internal DOT users and those with VPN access. STEP UP is available only to its pilot project participants. Ultimately, GeoMap2 will be made available to the public, although CDOT is currently determining which application (e.g., GoogleMaps, Microsoft Virtual Earth) could best support public access to GeoMap2.

The internal DOT departments that own the various datasets each set policies regarding their data but CDOT's GIS Section maintains and updates many data layers for GeoMap2. Other data layers are maintained and updated by data owners. FHWA monies for CDOT's yearly work plan

---

[5] The STEP UP Phase II report is available at: www.dot.state.co.us/publications/PDFFiles/step2.pdf

provide the funding for GeoMap2. There is growing interest in using the system and CDOT anticipates making GeoMap2 publically-accessible in the future.[6]

---

[6]Development of a comment interface feature—a mechanism for public stakeholders to add comments or suggest changes related to proposed projects—is anticipated within the next 1-2 years.

NORTH CAROLINA – NC ONEMAP

www.nconemap.com

CONTACTS
Alena Cook, North Carolina DOT (NCDOT)
Tim Johnson, CGIA
Jay McInnis, NCDOT
Dan Thomas, NCDOT

APPLICATION HOSTED BY:
North Carolina Center for Geographic Information and Analysis (CGIA)

KEY POINTS RELATED TO PEL

- Geospatial data clearinghouse for North Carolina.
- Incorporates 37 data layers on several priority data themes, including environmental and land cover themes.
- NC OneMap data layers are free for public download and available via a web-based viewer.
- North Carolina Center for Geographic Information and Analysis (CGIA) staff and the Interagency Leadership Team, a group of state and federal agencies involved in transportation planning and environmental decision-making, developed a business case to assess how a statewide geospatial data warehouse could contribute to cost- and time-savings for transportation agencies.

NC ONEMAP FOR PLANNING AND ENVIRONMENT LINKAGES

Development of NC OneMap

CGIA is the lead agency for GIS services and GIS coordination for North Carolina.[7] CGIA provides GIS services to state and local government agencies as well as the private sector and supports the North Carolina Geographic Information Coordinating Council (GICC), which develops statewide policies and standards regarding geographic information and GIS. The GICC coordinates NC OneMap as a collaborative process. A council of 33 members representing all NC OneMap stakeholders, including counties, the private sector, and state government, meets on a quarterly basis regularly to establish a consensus on which data layers to add to the system and policies regarding usage, access, and data-sharing. Some council subcommittees, which perform most of the detailed work, meet at least quarterly, while other working groups meet more often in between GICC meetings.

With the support of the CGIA, the GICC initiated NC OneMap, North Carolina's state geospatial data clearinghouse, in 2003 with the primary goal of making geospatial data accessible at a broader scale.

More than 90 of North Carolina's 100 counties use GIS to support a range of business needs, policy-related decision-making, and asset management activities. Prior to NC OneMap development, however, lack of a standardized data-sharing framework between counties or at the state level made it difficult for GIS end-users to have a comprehensive understanding of data availability. NC OneMap unifies geospatial data and makes a range of environmental information available to the public. OneMap's data set has statewide coverage for some data themes at some map scales, but has partial coverage for a number of other themes. The GICC is actively pursuing cost-sharing opportunities with State and Federal government agencies to ensure that

---

[7] North Carolina Center for Geographic Information and Analysis: www.cgia.state.nc.us/

data collection will be an ongoing effort. For example, current projects such as NC Stream Mapping Project,[8] a statewide stream-mapping initiative, will publish environmental data to NC OneMap once completed.

State legislative support for the program has increased over time and most recently, the North Carolina General Assembly created a database administrator position to manage NC OneMap content and an application developer position to support the program's web-based viewer and participant connections. Furthermore, there has been growing support for NC OneMap at the county level.

Assessing NC OneMap for PEL

The North Carolina Interagency Leadership Team, which is comprised of 11 state and federal agencies involved in transportation planning and environmental decision-making processes, developed a business case summary in 2006 to assess the cost-effectiveness for maintaining and developing a range of GIS data layers. Among other conclusions, the business case reported that enhancing and maintaining NC OneMap as a robust data warehouse would significantly benefit transportation and resource agencies. For example, the summary report found it likely that:

> Cost savings can be realized through decreasing the amount of overall time for project delivery through better screening of projects…and use of GIS to reduce the number of alternatives that need to be carried forward for detailed studies.[9]

The summary report also identified NC OneMap's potential benefit for facilitating project environmental screening, reducing delays to project implementation, and improving decision-making, all of which are PEL goals.

NC OneMap is a recent initiative and the potential benefits identified in the summary report are being realized. However, users recognized a few critical success factors that make NC OneMap an important data resource that furthers PEL. First, NC OneMap provides data consistency and multi-jurisdictional coverage, allowing users to efficiently access information for long-range planning. In addition, having access to standardized data has allowed users to work cooperatively, which has helped to expedite transportation project development processes.

Users view NC OneMap as a data repository that allows efficient 'one-stop shopping' for geospatial information. NC OneMap does not have a mechanism to track the history of data changes; however, it connects to county servers that provide the latest available information from that county to ensure that users access current information. Users report that NC OneMap data has improved the timeliness with which transportation projects and studies can be completed, since the OneMap viewer offers easy access to a large amount of data. For example, NC OneMap environmental data, including threatened and endangered species and historic data, was used to prepare a highway corridor study. The data allowed participating agencies to assess the study more comprehensively and efficiently.

NC OneMap staff and users have encountered several challenges in developing the data warehouse:

- Flow of Information – Users are not responsible for uploading data to CGIA: the flow of data from NC OneMap to users is generally a one-way process. Currently, there is no mechanism in place to allow users to update data while in the field or after completion of

---

[8] North Carolina Stream Mapping Project: www.ncstreams.org/
[9] Summary report of business case available at www.ncdot.org/programs/environment/development/interagency/NCILT/download/Goal1/GISBusiness_CaseReport.pdf or from NC OneMap upon request.

projects. Users who would like to continually update information to CGIA have found the lack of an appropriate mechanism to be challenging. GICC may consider adding a mechanism to allow this functionality in the future; however, before considering whether to add this mechanism, GICC would likely develop a pilot study to test how data updates from the field might be quality-checked. It is probable that GICC would rely on data stewards (e.g., NCDOT is the data steward for the roads data layer) to receive and quality-check data updates for content, completeness, and accuracy. The data stewards would then have the ability to either 'accept' or 'reject' data for posting to the live OneMap.

- Funding – NC OneMap relies on counties to obtain data. Information not collected by counties must be collected by the state; however, insufficient funding makes it difficult for the state to do so. In the past, catastrophic events, such as hurricanes or floods, have spurred funding for collecting environmental information. More reliable funding would help ensure consistent and ongoing data collection efforts. In addition, more reliable funding would increase the completeness of environmental data. According to a GIS Study Report prepared by the NC Office of State Budget and Management in February 2008,[10] NC OneMap land cover data layer is 80% complete, but lack of funding makes it difficult to increase data collection efforts.

- Ensuring Communication among Partners – It is challenging to develop NC OneMap at the statewide level. Some of the state's more rural counties, for example, may have only one GIS staff person or a small tax base. These counties may not have sufficient resources or time to support both county needs and initiatives, like NC OneMap, that are developed at the statewide scale. Urban counties are well-represented as NC OneMap partners but rural counties are less represented.

Features of NC OneMap

NC OneMap currently offers 37 data sets,[11] all of which are free to end-users and accessible via the internet. Data sets focus on priority themes, including land use/land cover, natural heritage areas, geologic features, water systems and supplies, and wetlands. Layers appearing in the viewer for a particular county, however, are not downloadable from NC OneMap, although this accessibility is a future goal. Counties providing data to NC OneMap decide whether these layers are free or are accessible for download; most counties make the data freely available from their websites.

Approximately 90 counties, cities, towns, state agencies, federal agencies, and regional councils of government are connected to NC OneMap. CGIA provides the data server and technical support for NC OneMap partners.

---

[10] GIS Study Report available at www.osbm.state.nc.us/files/pdf_files/GISStudyFinal02012008.pdf or from NC OneMap upon request.
[11] NC OneMap Implementation: Initial Data Layers to Serve: www.nconemap.com/Portals/7/documents/37NCOneMapDataLayers.pdf

OREGON – PORTLAND METRO REGIONAL LAND INFORMATION SYSTEM

www.oregonmetro.gov/maps

CONTACTS
Mark Bosworth, Portland Metro
John Mermin, Portland Metro
Dick Walker, Portland Metro
Matthew Hampton, Portland Metro

APPLICATION HOSTED BY:
Portland Metro MPO

KEY POINTS RELATED TO PEL

- The Regional Land Information System (RLIS) is a clearinghouse of regional geospatial data used for land use and transportation planning.
- Users include municipalities, counties, developers, and consultants; the public has free access to a web-mapping component of RLIS.
- The RLIS base map is at the parcel level and contains nearly 200 environmental and transportation data layers.
- Access to RLIS data is available on a subscription basis at several levels to member jurisdictions, other agencies, and consultants, but some RLIS data is available free of charge on the internet.
- RLIS is collaboratively managed.

RLIS FOR PLANNING AND ENVIRONMENT LINKAGES

Portland Metro MPO
Metro, the Portland area MPO, is the elected regional government for greater Portland, Oregon. Metro serves more than 1.4 million residents in 25 cities in the Portland region as well as Oregon's Clackamas, Multnomah and Washington counties. Portland Metro's mission is to provide a variety of regional planning and coordination solutions to address growth, infrastructure, and development issues that cross jurisdictional boundaries.[12]

Metro's Regional Land Information System (RLIS), which Metro maintains and hosts, is the major GIS for the greater Portland Metro region. With initial impetus from an executive champion within Metro, RLIS was developed in the mid-1980s as a data consortium to support a variety of land-use and transportation planning projects. RLIS has grown to now include nearly 200 GIS-based data layers that include sidewalks, bicycle routes, rivers, vegetation cover, parks, open spaces, sensitive habitat, wetlands, and historic properties that are linked to both street and census geography. In its role as a regional government agency, Metro maintains RLIS as a way to develop regionally-consistent land information and transportation modeling.[13]

Four full-time GIS technicians in Portland Metro's Data Resource Center maintain the RLIS and collect its source data. Managing RLIS, however, is a collaborative process that involves ongoing coordination with partner agencies. While Portland Metro does not earmark funds for RLIS outreach efforts, RLIS staff attend conferences and other events to market the system whenever possible.

---

[12] Portland Metro's Mission, charter, and code: www.oregonmetro.gov/index.cfm/go/by.web/id=24270
[13] GIS: Essential Technology for Urban Growth Management in the Portland, Oregon Metropolitan area (presentation by Richard Bolen, 2002) at: www.oregonmetro.gov/files/maps/gis_and_planning.pdf

RLIS' main users are local member jurisdictions and a variety of stakeholders ranging from business owners to municipal firefighters and land-use planners. There are approximately 120 external commercial clients, who access RLIS on a fee-for-service subscription basis. Internal users (e.g., Metro's Parks and Planning Department) draw on RLIS datasets to support a variety of business needs and projects. Internal users include the Parks Department, and the transportation modeling group and regional transportation planning and land-use planning sections. RLIS' user base is expanding as awareness of RLIS and its capabilities grows.

RLIS Support of PEL and Transportation Planning
Metro staff describe RLIS as a 'unique' data institution that presents comprehensive information on the transportation systems and urban environment of the greater Portland area. In doing so, RLIS presents information that facilitates environmental analyses for transportation planning.

RLIS has been used in a variety of ways to support transportation planning. For example:

- RLIS data were used to update Portland Metro's Regional Transportation Plan (RTP), which in part helped to identify levels of investment required for various transportation corridors.

- RLIS data populated maps for the RTP that overlaid sensitive U.S. Fish and Wildlife (USFWS) habitat areas with wildlife incident hotspots (e.g., fish passage barriers), flood plains, wetlands, and historic properties. These maps helped to identify sensitive environmental areas that could be of particular consideration for the RTP. While using RLIS data for the RTP update was not the first major use of RLIS data for transportation planning, it was the most in-depth use.

- RLIS is used to identify environmental factors for specific corridor projects.

- RLIS has supported development of MetroScope, Portland Metro's land-use modeling tool.[14] MetroScope uses RLIS data to predict future development scenarios for the Portland Metro area and make decisions related to urban growth, land allocation, and transportation planning.

- RLIS data has enabled travel forecasting as well as transportation travel time and traffic volume analysis.

Assessing RLIS for PEL
RLIS staff have not completed a formal cost-benefit analysis to assess the system, but there is a general consensus that RLIS has been very effective in producing products that are sensitive to both transportation decision-making and consideration of environmental factors. For example, RLIS data were used to develop a regional bike network for the Portland Metro area and an associated bike path map, titled 'Bike There,' which won an international award for map design.[15] In addition, Metro was honored with a Special Achievement in GIS award from the Environmental Systems Research Institute (ESRI) in Redlands, CA. The award recognized RLIS as "an essential tool for…environmental and natural resource management," among other applications.[16]

RLIS is an important data source for a broad range of decision-makers and stakeholders and general user response to RLIS has been very positive. One user, for example, commented that RLIS "is an absolutely vital factor" in his ability to integrate environmental data into transportation planning. This user stated that RLIS provides easy, 'one-stop shopping' access to data that he would otherwise need to obtain through other means.

---

[14] See also www.oregonmetro.gov/files/maps/gis_and_planning.pdf for more information on RLIS and MetroScope.
[15] MAPublisher Map Awards 2005: www.avenza.com/MPcomp/2005/
[16] ESRI International User Conference 2007: http://events2.esri.com/uc/2007/sag/list/?fa=Detail&SID=568

Several factors contributed to RLIS' success. As a centralized data repository, RLIS standardized data that had previously been derived from multiple sources and required standardized GIS practices from its consortium members. RLIS also offers reliable and robust data that are easily accessible and the system is regarded as a source of reliable, accurate, and current data. Finally, as one of the only data providers that has GIS-enabled data for the Portland Metro area, RLIS offers users easy access to information that would otherwise be more difficult to find.

Portland Metro staff noted several lessons learned that were important throughout RLIS' development. These lessons learned demonstrate that RLIS has broad applications to PEL in emphasizing early collaboration among transportation and resource agencies to streamline decision-making.

- Establish common ground – In order to establish a system that meets a broad range of needs, it was important to first establish common goals and practices among RLIS' primary stakeholders. In order to establish commonalities, RLIS staff worked with stakeholders in a series of meetings and workshops to establish a 'responsibility matrix' that outlined roles and responsibilities for data collection and maintenance in the region. Through these conversations, Metro worked collaboratively with stakeholders to choose the best standards for RLIS format and practices. Finally, RLIS staff compelled other users to meet those standards if they wanted to participate in the system.

- Be aware of possibility for technological change – RLIS staff understood that technological changes were possible, especially regarding how geospatial information was stored. To ensure that users were apprised of new technologies, RLIS staff maintained training and outreach efforts among all users as well as participating transportation and resource agencies.

While RLIS has been successful in linking environmental consideration to transportation planning, staff have encountered some challenges in developing the system:

- Maintenance – RLIS staff report that it is challenging to determine how data should be updated, maintained, and funded. For example, staff teamed with a transit agency in the early 2000s to obtain sidewalk data for the Portland Metro region, but those data have not been updated. It has been difficult to determine how to balance the need for obtaining accurate, current data with the realities of data cost.

- Business model – RLIS data is available to consultants and developers for an annual fee of $1,000. Portland Metro has been occasionally encouraged by citizens to make the information free of charge because of the Freedom of Information Act. However, Portland Metro believes that RLIS provides value-added over and above the actual data and refers FOIA requests to the relevant jurisdiction or other agency that collected the data. The data is accessible for free, but not in the RLIS-GIS context.

- Establishing standards – An ongoing challenge is maintaining and providing data effectively in the context of a consortium of multiple jurisdictions and agencies, some of which operate only within municipal boundaries and others (notably the transit authority) which operate throughout the region. When Portland Metro first created RLIS, it was necessary to bring these jurisdictions to agreement on standardizing roles and technical practices.

Data Sources, Funding, and User Access
RLIS data were initially derived from information available on the U.S. Census Bureau's Topologically Integrated Geographic Encoding and Referencing System (TIGER) system and other private organizations such as the Pacific Gas & Electric Company (PG&E). Now, RLIS has several data providers who update the information on a quarterly basis. Data are collected from

local jurisdictions and other agencies such as the U.S. Census Bureau, Bureau of Labor Statistics, and Bureau of Economic Analysis. While there is no formal agreement in place between RLIS and ODOT, data are shared and exchanged with both the DOT and Oregon's statewide data clearinghouse, the Oregon Geospatial Enterprise Office's (GEO) Spatial Data Library,[17] on an informal but regular basis.

The base map for RLIS is at the parcel level, which is provided by the local county tax assessors. Public access to the system is obtained through paid subscriptions. There is, however, a substantial web-mapping component to RLIS that is free and accessible to the public. Subscription revenues constitute a small percentage (approximately 10 percent) of the total funding required to support RLIS. Metro general funds and Federal grants for transportation planning are the primary funding sources.

---

[17] Oregon Geospatial Enterprise Office Spatial Data Library: www.oregon.gov/DAS/EISPD/GEO/alphalist.shtml

III. CASE STUDY FINDINGS AND CONCLUSIONS

FINDINGS

The following section summarizes the key findings that emerged from the case study interviews. Key findings include:

- Agencies developing GIS for PEL may take different paths.
- Interagency collaboration is essential for data-sharing.
- Funding availability and policies may vary.
- Data are acquired from different sources, necessitating standardization.
- Data confidentiality policies differ.

KEY FINDING 1: AGENCIES DEVELOPING GIS FOR PEL MAY TAKE DIFFERENT PATHS

The three organizations hosting the case study applications followed different paths to developing the technologies. Figure 1 summarizes and compares the different choices that the three case study agencies made in developing GIS.

Figure 1: Primary Factors for Using GIS to Implement PEL

PRIMARY FACTORS	GEOMAP2	NC ONEMAP	RLIS
Business model	• Desktop environmental application hosted by State DOT	• Free state geospatial data clearinghouse hosted by state GIS agency	• Subscription-based data clearinghouse hosted by MPO
Funding model	• FHWA-funded	• Cost-sharing between the Geographic Information Coordinating Council and state and federal government agencies	• Federal grants • Annual dues from member jurisdictions • Fee-for-service subscriptions
Users	• CDOT • Resource agencies	• State agencies, including NCDOT • Public (no current limits on access)	• Member jurisdictions • Commercial clients such as consultants and developers (fee-for-service access) • Public (limited free access)
Uses	• Internal DOT project planning • Enhancement of an earlier internal desktop application for environmental data	• Improve environmental decision-making at project development level • Environmental screening for planning projects • Land use planning	• Transportation planning • Project-specific traffic modeling • Land use planning
Data sources	• Highway data and a desktop application (data from the desktop application are being	• County GIS data sets • Purchased data sets	• Purchased data (imagery and photography) • Project-specific

15

	transferred to a database that will be used in GeoMaps2) • Previous databases such as STEP UP		reports from specific projects • Local jurisdictions add data
Data confidentiality	• Currently serving confidential data.	• Not currently providing confidential data.	• No confidential data in system

In addition to the factors listed in Figure 1, other unique factors that influenced the scope of the GIS applications were the occurrence of extreme weather events (in some cases, these occurrences appeared to motivate public calls for better environmental mapping) and development of business cases to assess the cost- and time-benefits that the GIS provided.

KEY FINDING 2: INTERAGENCY COLLABORATION IS ESSENTIAL FOR DATA-SHARING
In creating their systems, the three agencies hosting the case study applications invited different levels of participation from resource and transportation agencies and other stakeholders before making technical and policy decisions about the software platform and included data. The degree of collaboration the application-hosting agencies sought was directly related to the intended pool of users. For example, in North Carolina, where the GIS was intended to be open to all state agencies, collaborating agencies jointly made policy decisions with their respective user bases. Portland Metro and CDOT were able to work with future partners to identify common needs and interests and then were able to make policy decisions internally.

The degree of collaboration obtained by the three case study agencies also appeared to be influenced by stakeholders' concerns about whether the system could meet their own business needs. More specifically, when data was sought from another entity, such as a county administration or a resource agency, some of the case study agencies found it difficult to obtain permission to use information from the data owners. Data owners sometimes reported that they did not have the necessary time or staff to meet such requests. Both North Carolina and Portland Metro found it challenging to build a strong enough case to obtain cooperation from stakeholders to share data.

KEY FINDING 3: FUNDING AVAILABILITY AND POLICIES VARIED
Funding was a limiting factor for some of the three case study agencies but not all. In addition, financing options in some locations were restricted by state policies or statutes.

The application-hosting agency's funding needs appeared to have several components, which included initial development costs, operational costs, and the cost of acquiring and updating data. In particular, data owners (such as resource agencies or county governments) sometimes required funding to provide data or cover staff costs for the effort.

For Portland Metro and the North Carolina Geographic Information Coordinating Council (GICC), it was necessary to convince others of the long-term efficiencies and effectiveness of GIS in order to obtain political and/or financial support. This required quantifying the cost-savings for all potential GIS users and making a case for substantial up-front and ongoing investment in GIS.

KEY FINDING 4: DATA ARE ACQUIRED FROM DIFFERENT SOURCES, NECESSITATING STANDARDIZATION
Data were commonly acquired from many different sources, necessitating a common format. For example, data were obtained from existing GIS databases at local jurisdictions, resource agencies' maps or databases, special studies associated with project-specific NEPA documentation, federal databases, by fee from private consultants and photography services, or from the application-hosting agency field staff using hand-held GPS. Data scale/level of detail, accuracy, and frequency of updates varied widely among the interviewed agencies.

There were several considerations that appeared to influence application-hosting agencies' choice of software platform, including the level of data detail, the degree of data accuracy, and whether the data were up-to-date. For example, site-specific data from a private study may be accurate to within a foot while landscape-scale data from a public database may be accurate to within a few tens of feet.

KEY FINDING 5: DATA CONFIDENTIALITY POLICIES DIFFER

While data confidentiality policies differed, transportation agencies appear to have a common need to access some confidential information, including locations of sensitive resources, in order to streamline environmental review processes and communicate effectively with resource agencies. However, concerns about network integrity and other political and funding considerations affected whether data were shared only internally within the application-hosting agency or with other agencies and/or the public, consultants, or developers.

CHALLENGES

Both the interviews and information acquired during the literature review showed that developing and implementing a GIS application to link environmental and transportation planning concerns is a challenging, but ultimately useful, enterprise. In all three cases, agencies encountered major challenges that arose when the GIS application was extended beyond the boundaries of a single agency.

Some other common challenges reported by the case study agencies included:

- Finding reliable funding, particularly for ongoing operations;
- Coordinating across multiple agencies to agree to the same platform and format for obtaining and sharing data;
- Obtaining and maintaining accurate and comprehensive data sets; and
- Addressing the need to standardize data formats.

LESSONS LEARNED

The case studies demonstrate that GIS can be effectively used to implement PEL in a variety of ways. Measures for effectiveness varied, but interview contacts generally used qualitative assessments to report on the ways in which GIS had increased efficiency, collaboration, or environmental considerations in transportation planning processes. For example, North Carolina's NC OneMap allows end-users to more efficiently access a range of environmental information, reducing screening delays for transportation planning. Colorado's GeoMap2 will integrate statewide environmental data with data already collected at the MPO-scale to enhance internal planning projects. Finally, Portland Metro's RLIS has provided users with accurate environmental data to use in regional transportation plans, corridor studies, and land use modeling to facilitate public acceptance and make better decisions.

One exception to the use of qualitative assessments was the Interagency Leadership Team (ILT) in North Carolina, a group of federal and state agencies involved in transportation and environmental planning, which developed a business case to quantitatively evaluate the benefits of maintaining statewide GIS data layers (as represented by NC OneMap).[18]

---

[18] The business case was developed by the Interagency Leadership Team in North Carolina and is discussed in more detail in the NC OneMap case study in Chapter II of this report. The summary of the business case is available for download at www.ncdot.org/programs/environment/development/interagency/NCILT/download/Goal1/GISBusiness_CaseReport.pdf

Whether effectiveness was measured qualitatively or quantitatively, several factors appeared particularly important to ensure effective application of GIS to PEL. These factors have been identified from the interviews to capture a few general lessons learned:

- **GAIN AGREEMENT ON DATA PLATFORM AND DATA-SHARING GOALS** – Several contacts identified issues involved with standardizing software and the ways in which data were collected, managed, and stored between counties and the state, or between transportation and resource agencies. These contacts reported that establishing agreement on data platforms and data-sharing goals was an important step to take in developing GIS that met the needs of all end-users. In addition, contacts reported difficulties in standardizing the level of detail for data collected from counties, the state, or other agencies. Another lesson learned was to establish the most appropriate level of detail required by system users.

- **CULTIVATE A CHAMPION** – Some contacts reported that cultivating a champion could help to motivate GIS use across an agency. Specifically, a champion helped overcome resistance to losing control of specific data sets. In addition, a champion brought broader perspectives to a project when its initial focus was narrower. For example, in the case of Portland Metro's RLIS, an executive-level champion helped to build the program from a local data consortium to a multi-jurisdictional information clearinghouse.

- **COLLABORATE WITH STAKEHOLDERS** – Close collaboration with stakeholders during the planning phase and throughout the life of the project was another important lesson learned. Collaboration occurred through ongoing meetings that brought data-producers and end-users together to discuss information access, collection, and usage policies. Collaboration also occurred while working with partners to transition environmental data from small-scale GIS applications to expanded ones and developing business cases that assessed GIS cost- and time-effectiveness.

IV. CONCLUSIONS

The FHWA PEL program has supported use of GIS as an efficient and effective tool for integrating environmental considerations into transportation planning and project development. The three case studies presented in this report illustrate several specific challenges to agencies setting up or expanding GIS systems that link environmental and transportation data. Some of these challenges are common; others are situational.

Taken together, the case studies indicate that there are diverse business models for how GIS can implement PEL. The applications detailed in this report varied from a statewide geospatial data warehouse to a DOT-hosted internal desktop application as well as a subscription-based statewide data clearinghouse. While these business models varied, contacts reported several key considerations that suggest some best practices and lessons learned for addressing PEL goals through GIS. These considerations, which are outlined in more detail in Chapter III, include addressing data sources, uses, and funding models to allow a common understanding of GIS objectives and end-users' needs.

V. APPENDIX A: INTERVIEW QUESTIONS

PROJECT HISTORY

- How/why did the project begin?
- What division/section/agency initiated it? Did it need/have a 'champion'?
- What was the motivation? What were the goals?
- Were there obstacles? What were they and how did you overcome them?
- When did development begin and when did the system come on-line?
- Has it been used for planning? For specific projects? (more questions on this below)
- Is development completed? What changes or additions are planned, and when?

WHAT IS THE ORGANIZATIONAL CONTEXT OF THE PROJECT?

- What division or section or agency manages it?
- Has the 'location' of the project (i.e., within a division, agency, or outside the DOT) been a factor contributing to successes/challenges?
- If "located" or managed outside the DOT, how has the DOT been involved with this project?
- What divisions/agencies are active participants in using, maintaining, or championing continued use & development of ___?
- Who determines policies for data management? What is chain of command for additions or modifications?
- What agreements, if any, are in place to ensure confidentiality of sensitive data (e.g., location of rare ecological or cultural resources)?

FUNDING

- How is ___ funded now? Same source(s) for maintenance and for development?
- Have any cost savings been identified and/or quantified from streamlining of environmental review?

DATA ACQUISITION

- How many layers are included? What information is contained in layers?
- How were/are data for ___ acquired? (bought, collected in field, mined from internet, donated)

ACCESS

- Who has access to data in the GIS application? Do all users have the same level of access? Can all users access the same data layers?
- How do users access ___?

TECHNOLOGY & FUNCTION

- What is the technological context of the application?
- Does your agency run other GIS applications?
- Does ___ interact with those applications? How?
- Does a transportation planner have to export data to do transportation planning? Modeling? What feedback do you get from them?
- How do you identify benefits of this project to users?
- What is ___ primarily used for at this time? By whom?
- What has been the response from users/stakeholders? Do you have a feedback mechanism?
- How is feedback from users managed/communicated to project decision-makers?

- Can you describe past/current feedback from users?
- What do you believe are the best features of ___? Why?
- What limitations does __ have, in your mind? Are these likely to be changed?

ENVIRONMENTAL CONSIDERATIONS IN TRANSPORTATION PLANNING

Role of ___ in facilitating early considering of environmental issues

- Does ___ help facilitate environmental stewardship/streamlining goals early in the planning and development process? If so, how? Can you provide examples?
- Does ___ facilitate public involvement/collaboration? If so, how? Examples?
- Does ___ improve decision-making? If so, how? Examples?

Success in streamlining NEPA

- Has ___ been successful in meeting the goal to streamline the NEPA process?
- How do you define "success"? How is "success" measured? Cost-benefit analysis, etc.?
- Has the program created significant time-savings and/or cost-savings?

Best practices & transferability

- What would you consider to be the "best practices" and/or "lessons learned" from ___?
- What would you recommend to other agencies that are considering use of GIS for streamlining environmental considerations of projects?
- What parts of ___ are duplicative? Do you think your GIS could have national transferability? Why/why not?

OTHER

Suggestions

- Any more information you would like to add?
- Any other documents/literature we could review for more information about the application?
- Can you recommend any other contacts for learning more about ___?

www.ingramcontent.com/pod-product-compliance
Lightning Source LLC
Chambersburg PA
CBHW081824170526
45167CB00008B/3528